Structural Retrofit – A Tangential Touch

(Applied knowledge of structural retrofit and allied topics.)

AF660640

Er. Chirag K Baxi. and Er. Kalpesh D Savaliya

"Connect the Vision"
Nexus Stories Publication
Bharata

NEXUS STORIES PUBLICATION®
Surat, Gujarat, India.

Title: Structural Retrofit - A Tangential Touch

First Published by Nexus Stories Publication 2025

Copyright © Er. Chirag K Baxi. and Er. Kalpesh D Savaliya
All Rights Reserved.

ISBN # 978-93-48504-71-5

This book has been published with all efforts taken to make the material error-free after the consent of the author. However, the author and the publisher do not assume and hereby disclaim any liability to any party for any loss, damage, or disruption caused by errors or omissions, whether such errors or omissions result from negligence, accident, or any other cause.

No part of this book may be used or reproduced in any manner whatsoever without written permission from the author except in the case of brief quotations embodied in critical articles and reviews.

Declaration: This book's content is owned by the author/editor. Nexus Stories Publication™ is a service provider for book design/distribution.

Publication
Nexus Stories Publication®, Surat (Gujarat), Bhārata
https://nexus-stories.com # +91 87800 80718

Author's Space ....

Retrofitting or refurbishing or rehabilitating the damaged structures is nothing less than giving it a new life after it has arrived at the threshold of dying. It is far complicated than constructing a new structure under strict quality control and very specific design drawings.

An important factor this action is that every damage needs to be individually analyzed, assessed and evaluated to design a tailor-made retrofit scheme and method so that the success is near 100%.

Few factors which are essential for designing and executing the retrofit scheme are addressed in the following pages to help one and all who are engaged in designing and executing the structural retrofit of damaged concrete and steel structures.

Addressing those necessary aspects which are absolutely necessary for a structural retrofit action all throughout the concept to creation stages; makes this handbook unique and important.

Hope, it is useful to you.
BEST LUCK .....

- **Chirag K. Baxi and Kalpesh D. Savaliya**

Preface

Ashok Khurana

Director General (Retd.), CPWD, Govt. of Indi a
Former Engineer Member, DDA

Preface

Concrete corrosion refers to the degradation of concrete due to the exposure to certain corrosive substances usually sulphur compounds, salty or acidic water, microbes, chlorides, nitrates or fluorides.

Concrete corrosion is the process by which concrete structures and components deteriorate due to chemical, colloidal, or physicochemical reactions. This process can cause widespread damage to the concrete structures like bridges, piers, pipelines, buildings.

In India and other developing countries corrosion of concrete and damage to the structures it's not uncommon. To restore the structure to its original strength or to enhance strength, requires structural retrofit.

Structural retrofit is a scientific activity to be carried out holistically for the purpose of enhancing the functional life of the underperforming structure. One has to look at structural retrofit at macro level as well as micro level, such as joints of the structure, indicative weak points even though not structurally damaged must be taken care of.

In this book the authors have talked about measures to prevent damage to concrete, causes, visual inspection and reasons for structural damage.Dos and don'ts of structural rehabilitation, importance of structural design and dynamics in rehabilitating the damaged structures. They have also talked about for protective anticorrosive measures for concrete and steel members.

I appreciate the authors for elaborately discussed the above points in this bookin a simplelanguagebut in-depthanalysis has been done. They have rightly titled it "Structural Retrofit a Tangential Touch".

FIE; Member: IRC, IBC, IWWA, ICAI, IIT Arb. CEAI; Chartered Engineer, Arbitrator, Conciliator

L-109, Sarita Vihar, New Delhi-110076, **Mob.:** + 91 8800690910, Ph.: 011 41090910
e-mail: ashok.khurana1@gmail.com

It is a helpful book for all thosewho areinvolved in structural retrofit activities in general to go through this book to understand the whole scenario and activities that arerequired to be designed and taken up for the retrofit, and also design component wise specific requirements.

Ashok Khurana

TABLE OF CONTENTS

Measures to prevent damage to concrete

It is beyond doubt that concrete is a versatile substance on the earth having its place in first five items being consumed on the earth. It is a well-crafted entity by the age-old civil engineers to provide flexibility, comfort and desired parameters to the constructed elements including those which are under construction. It is a matter of significance for any civil and construction engineer that nearly 8 trillion MT of concrete of different categories is manufactured in the world every year.

However, concrete has to be manufactured with strict quality control for it to perform as desired. Standards set in different countries all cross the world including India govern its quality parameters which broadly cove the following.

- Cement content:

Each grade of concrete, demands minimum cement content for ensuring proper binding and ultimate integrity of concrete matrix. Codes prevailing in different parts of the world have specified this value. This has been specified as 360 Kg/Cubic Meter of concrete in IS 456 under Indian Standard Code. It is further fine tuned under concrete mix design which is carried out based on sieve analysis of course and fine aggregates to be used to manufacture concrete. It goes up to as high as 425 Kg/Cubic meter for achieving high strength concrete. However minimum cement content is a basic criterion which should be invariably maintained.

- Slump:

Rheology of concrete is measured by slump test. This is a =n optimum value of workability of concrete which should be just in line with the specified limits under standard code applicable in respective part of the world. Although it is related to climatic and atmospheric conditions of the country, its minimum and maximum values must be holistically adhered to.

- Density:

One of the most important parameters contributing to sustainability and durability of concrete is its density. Normal density of concrete is 2,400.00 Kg/Cubic Meter. This must be essentially maintained to minimize the porosity of concrete matrix. It is a fact that higher porosity gives an easy access to the external contaminants to enter the concrete mass and deteriorate the same including reinforcement.

- Water permeability:

This is very significant parameter which is directly related with life of any concrete structure. Water penetration within concrete mass results in reinforcement corrosion which is a threat to the functional life of concrete structure. Quality of water is directly proportional to the extent of damage which the water permeability can cause to concrete structure. Salty or saline water can bring quick deterioration of reinforcement and concrete mass.

- Clear cover:

This is also one of the specified parameters of standard concrete under Code of any country. Clear cover of concrete over reinforcement relates with type of member. It is different for slabs, beams, columns and foundations. It is also influenced by the surrounding conditions prevailing around the concrete element. In case of an inadequate clear cover for concrete, the external contaminants can easily approach reinforcement steel to react with it and corrode it.

Causes of concrete corrosion.

1. High humidity,
2. High temperature,
3. Saline conditions,
4. Reinforcement corrosion,
5. Design inadequacy,
6. Chemical environment (gaseous, liquid or solid chemicals coming in contact with concrete).
7. Bird Dropping.
8. Substandard construction practices.
9. Forced compromises because of Financial control.
10. Lack of maintenance of concrete structure.
11. Failure of concealed services.
12. Use of improper construction materials for respective function.
13. Unforeseen natural calamities like earthquake, blast, flood, fire etc.
14. Stagnation of corrosive chemicals.

Visual inspection for structural damage / condition assessment.

Visual inspection and structural damage / condition assessment are two sides of a coin called structural rehabilitation.

Visual inspection is to be very carefully, critically and scrupulously carried out.

Visual inspection lays a foundation of structural damage / condition assessment. Inspection of the surface under address has to be careful and close. Spalling of concrete, deflection of concrete member, cracks, signs of fungus, differential (horizontal or vertical) movement of the member, traces of rust colour on the surface, bulging of member etc are few of many signs of structural damage or deficiency. Width of crack, degree and distance of deflection, buckling width of member, surrounding conditions etc can be found out only with careful and critical visual inspection.

Reason of each defect / deficiency / damage are to be individually evaluated and identified. This is possible only with close and careful visual inspection because it forms a base of further investigation like tests (non-destructive or semi-destructive) to have better idea about root cause of damage. It fairly helps in distress mapping of the member which can lead towards software analysis of the damages to arrive close to identification of particular parameter with its magnitude of loss.

Just as many zeros carry the value only after they follow the 'figure one', visual inspection is that 'figure one' for all the exercises of structural damage / condition assessment.

It is hence mandatory that the first and foremost step towards structural damage / condition assessment - which is visual inspection - is carried out by an experienced, knowledgeable and an expert technocrat to take the further investigation in the most appropriate direction.

Reasons for structural damage – Part 1:

Concrete and steel structures get corroded and damaged due to many reasons. Each reason has its significance and gravity attached with it. An effort is being made in this series of notes to address each reason in adequate depth.

The first reason is construction materials / ingredients of inferior quality.

Every part of the world has defined / specified quality of construction materials / ingredients for achieving desired quality of the constructed structure.

If quality of any of the materials is inferior to the quality as per applicable standard, it has direct effect on final / ultimate quality of the structure.

Poor quality of water used for construction carries contaminants like chlorides, sulphates and other dissolved impurities. They have direct effect on cement hydration process. Finally, the hydration takes place in presence of those contaminants. This means that the process of gain of designed parameters of the structure, happens in presence of the contaminants which inculcates defects and deficiencies right at the time of birth of the structure.

Fine aggregates contribute towards plugging the interspace of coarse aggregates. If they are containing silt or other impurities, the actual quantum of fine aggregates (of desired quality) available within concrete mass would be less than the calculated quantity as per mix design because volume of fine aggregates added to concrete would also contain silt and other impurities –

which would also carry their volume. Accordingly available quantity of fine aggregates within concrete mass becomes less than the designed quantity and presence of contaminants with fine aggregates would restrict gain of designed parameters. Silt and other impurities don't contribute to strength gain process.

The function of coarse aggregates is to provide strength to the concrete matrix. Coarse aggregates with fractured texture or flaky / elongated shape, don't contribute designed strength to the concrete matrix. Awkward shape of coarse aggregates, infringe development of cohesive concrete matrix which is a significant sign of developing deficient concrete.

Thus, concrete member which is constructed with construction materials of inferior quality have tendency of giving rise to defective concrete which would require continuous monitoring of parameters followed by corrective actions.

Reinforcement steel is one of the major construction ingredients and its quality is also to be addressed with adequate focus. This will be discussed in next note.

Reasons for structural damage – Part 2:

The subject was deliberated in my last note with focus on role of construction materials for deterioration of concrete members. As said therein, this note is focussed on role of reinforcement steel as reason for structural damages.

Fresh reinforcement steel for a RCC member has to be as per prevailing and applicable standards. Physical properties and chemical composition of steel are well defined in respective standard which have to be strictly followed.

Reinforcement steel supports tensile load. It's property of ductility is also very significant.

If reinforcement steel has its properties at variance from the specified mechanical properties and chemical composition, then that reinforcement steel might fall short of supporting tensile force and deflection resistance - required under lateral thrust suffered by the member even without any effects of corrosion to reinforcement steel.

Another check for reinforcement steel is its manufacturing status. If it is rolled steel or re-rolled steel, it is definitely prone to inferior chemical composition than the original and specified steel as per standard. This is because repeated melting steel at very high temperature has tendency to weaken the metallurgical properties of the original steel. It can either make it brittle or may reduce its corrosion resistance or might affect its ductility property etc.

Hence chemical composition, physical parameters and metallurgical matrix of reinforcement steel have to be ensured to be as per the prevailing and applicable standard.

Reasons for structural damage Part 3:

Concrete and steel structures get corroded and damaged due to many reasons. Each reason has its significance and gravity attached with it. An effort is being made in this series of notes to address each reason in adequate depth.

One of the important reasons is High humidity and high temperature.

Concrete is very strong and versatile but has few significant limitations which have to be addressed while creating a concrete structure. Material criteria (aggregates, water and reinforcement steel) have been discussed in earlier two notes. It is intended to address climatical challenges for concrete structures in this note.

High temperature affects hydration process of concrete. Exotherm heat of cement hydration needs to be compensated by cooling by external means or otherwise. If temperature of climate surrounding concrete is higher than the specified temperature and the curing process of concrete create differential thermal conditions. High temperature expands the object and lower temperature contracts the object. Accordingly, the concrete member exposed to high temperature during construction or during its service life often tends to develop differential expansion resulting in cracks of different types - most of them being structural cracks.

If high temperature is accompanied with high humidity, it aggravates the differential movement of the concrete member. Concrete breathing phenomena is obstructed under combined high temperature and humid conditions.

Results of combined effects of high temperature and high humidity on concrete member is development of specific stresses for which the member is not usually designed. More often than not, the process of cracking happens because of reinforcement corrosion but here, though reinforcement steel is not found corroded, the member could get buckled or deflected.

Suitable protective and breathable coating around concrete member is an appropriate solution to this category of limitation for concrete member.

Reasons for structural damage Part 4:

Structures are often found damaged because of many reasons. Effect of reason for damage depends on impact of that reason and inherent strength of the structure. Different reasons connected with ingredient materials and climatical challenges have been discussed in earlier three notes. It is aimed to address a sensitive reason in this note which is related to quality of construction.

Few of common mistakes during and after construction activity which can be dangerous at later date causing damage to concrete are listed below.

1. **Incorrect Water : Cement ratio**. It is often tampered to adjust workability of concrete.
2. **Incorrect method of concreting**. Height from where concrete is to be placed in formwork should not be more than 1.25 meter. Even over-vibration can cause segregation.
3. **Poor formwork**. Formwork which is not watertight will allow cement slurry to flow out of concrete matrix which will reduce availability of cement for polymerization.
4. **Inadequate curing**. Loss of thermal equilibrium for fresh concrete matrix is very dangerous.
5. **Insufficient curing**. Exposing the concrete member to loads develops differential stresses which aggravate at later date causing structural damage.
6. **Design compromise**. Insufficient cement content or steel content for the concrete member due to any reason is one of the most serious threats for a concrete member.

7. **Incorrect or no coating over surface after curing**. Once the curing time is complete, the member has to be coated with compatible coating to protect it from the weather or chemical environment and also to seal the surface porosity to prevent any contaminants to enter the core mass through surface porosity.

These are few of many factors which affect quality of concrete member right from the time of its construction. If these mistakes can be prevented and avoided by necessary care, then the constructed concrete member would be of desired quality and would successfully perform throughout its designed functional life.

Let us aim at getting this quality of concrete.

Effects of climate and weather on structural rehabilitation activity:

When a damaged or underperforming structure is taken up for its structural rehabilitation / retrofitting / strengthening, its execution requires to be planned considering surrounding climate and weather conditions. It is briefly addressed below.

1. Ideally atmospheric temperature during such activities should be in the range of 27^0 C to 35^0 C. This is significant especially when the activity involves repair material having high exothermic polymerisation process. Lower atmospheric temperature than 27^0 C or higher than 35^0 C, has tendency to disturb thermal equilibrium between temperature of rehabilitation mass and atmospheric temperature.
   - 1.1 Use of ice is one of the essentials for taking up of these activities under high temperature.
   - 1.2 Concealing of conduits carrying cold water within the concrete matrix (by taking it into consideration right from design stage) and continuously blowing cold water throughout the activity of placing of micro-concrete; helps in minimising the adverse effects due to high temperature within the matrix as well as outside temperature.
2. Atmospheric humidity also affects the velocity of polymerisation process. Polymerisation slows down during high humidity and hence rehabilitation activity can be scheduled when atmospheric humidity is normal as per the meteorological data. When resinous materials are involved in structural

rehabilitation activity, polymerisation of its different ingredients drastically slows down under high humidity.

3. However, if it becomes absolutely necessary to execute the activity during high humidity, then the proportioning of water: powder ratio would need to be adjusted as per manufacturer's advice.

It is a standard practice of keeping a close watch on surrounding atmospheric conditions affecting structural rehabilitation / retrofit / strengthening activities to execute them with all precautionary measures so that the rehabilitated structure performs as desired.

Humidity in structures – a challenging threat.

Cementations structures are often found with (excessive) hidden humidity even after the curing period is complete. This is a serious factor affecting the functional performance of that structure or member.

1. It affects binding action and hence integrity of the concrete matrix.
   Integrity of a member is of utmost importance for its long functional life. Hidden humidity within the member affects hard binding action of concrete ingredients and then the integrity deteriorates.

2. It attacks reinforcement steel and corrodes it.
   Reinforcement steel exposed to humidity at a close distance can easily corrode fast.

3. It affects the stability of protective or decorative resinous coating applied on the surface of structure or member. It also obstructs the process of development of bond between the mother surface and any other treatment to be applied on the surface of structure or member.
   Any resinous protective or

decorative coating or treatment must be applied on the member only after its internal humidity has been completely evaporated.

4. It affects the plans to modify or refurbish the structure or member to take additional load or for change of the functionality of structure or member. Members having hidden humidity for any reason for a long time is normally not found fit for its structural or functional modification. This becomes a big challenge especially for industries.

Important tips for protecting structures in coastal region

1. Saline conditions causing structural damages in coastal areas is major point of concern.
2. Speed of construction with more thrust on cosmetic value at times induce the inherent defects which can spoil the structure with passage of time.
3. Absence of special and specific measures to plug / prevent sub-soil water entry to the foundations or basements of the structure is responsible for water ingress at that level which is beginning of end.
4. Exterior coating for cosmetic and aesthetic value to the structures and surrounding area must have element to resist saline atmosphere in addition to UV resistance
5. Use of GFRP bars as structural reinforcement must be extensively explored for construction projects. GFRP bars as a replacement of steel reinforcement bars is gaining momentum these days with distinct and significant advantages exhibited by GFRP bars. ACI CODE-440.11-22: Building Code Requirements for Structural Concrete Reinforced with Glass Fiber-Reinforced Polymer (GFRP) Bars is the best guide for selecting and deciding about use of GFRP bars as reinforcement for RCC member.

Chemistry of structural damage exposed to saline conditions

Concrete and steel structures exposed to saline and salty environment, get damaged and they are structural damages. They cause chaos at later stage. It is necessary to understand chemistry of these damages to arrive at the perfect rehabilitation method for them.

Concrete matrix essentially consists of coarse aggregates, fine aggregates, cement and water. Admixtures are added to the matrix as per the functional need. The concrete matrix exposed to saline environment, gets exposed to the salts present in that medium (liquid or solid or gaseous state). More often than not, Chlorides cover the maximum share in contents of saline medium. Chloride present in saline environment is mostly in unsaturated form and tends to quickly react with the molecule of chemical with befitting characteristics.

Alumina as present in cement which is prime contributor for imparting binding properties to cement in concrete matrix is one of the most easily vulnerable elements whose molecule attracts Chloride ions and gets reacted to form Aluminium Chloride. This form of Aluminium doesn't contribute to the binding property of cement and accordingly the homogeneity or integrity of concrete matrix which was provided by Alumina of cement gets lost and the matrix becomes heterogeneous. Integrity of the matrix reduces as a result of this process and that gives rise to increase in porosity, water absorption, corrosive chemical ingress etc. These are the signs of structural degradation and deterioration.

The most important step to protect concrete members exposed to saline or salty environment, is to provide the most suitable coating on them for not only protecting them against salt attack but also to protect from other factors which might be accelerating or aggravating the deterioration process. Exposure to saline conditions in interruptions, makes it more difficult for concrete to sustain this condition.

Myth prevailing about structural rehabilitation of concrete members near sea shore.

It is often believed that concrete structures are seriously and adversely affected by humidity and salinity of sea. This is true to some extent but the matter needs to be well understood.

The concrete structures which are continuously facing sea; are exposed to salinity and humidity to which concrete is not very resistant. Some inherent deficiencies in concrete member which might have developed or left out at the time of its construction like insufficient cover, insufficient compaction / vibration, lack of surface protection etc have their respective detrimental effects with higher intensity on concrete member which is continuously exposed to sea.

It is difficult to rectify the deficiencies left out during construction unless the member is taken up for its systematic and scientifically designed structural rehabilitation but surface protection of the member is a subsequent activity which must be holistically carried out. With innovative researches taking place in surface corrosion protection subject, many optional treatments are available out of which the most suitable option can be selected and surface corrosion protection treatment can be applied at an appropriate time (like curing period) after concreting so that a step forward can be achieved in reducing the possibilities of structural damage to the concrete member. It is strongly recommended to apply / implement such treatment even for the members which are taken up for their structural rehabilitation.

There is one more important point to be addressed for deciding the treatment for structural rehabilitation of concrete members facing sea.

Concrete members which are always remaining submerged, have less tendency of deterioration due to corrosion. This is because they don't get Oxygen for corrosion process at intervals. Unlike this, the members which are alternatively get submerged under circumstances like tidal variation, they get Oxygen - although interruptedly. That period of exposure to Oxygen is adequate for the member for undergoing corrosion process.

Hence while taking up structural rehabilitation of concrete members which are continuously facing sea, such facts must be discriminately addressed to achieve the ultimate success to the act of rehabilitation.

Challenges for civil structures for survival in Oil, Gas, Petrochemical, Chemical, Fertilizer Agrochemical and allied Industries.

Civil structures housing operational units, piping, equipment, pumps etc in Oil, Gas, Petrochemical, Chemical, Fertilizer Agrochemical and allied Industries, often face aggressively corrosive chemicals, solvents, hydro-carbons etc to which concrete and steel can't resistant. It becomes a significant challenge to ensure sound and strong parameters of cementations and steel structures in these industries.

The actions in this direction begin with protecting them from getting affected with the corrosive attack of the chemicals and other contaminants. Tailor made treatments to protect the structures from each contaminant or combination of contaminants with different damage potential, are to be selected and applied so that their structural health is maintained and they don't deteriorate when exposed to corrosive chemicals. This is extremely important because one failure in protection treatment, can allow corrosive chemical to penetrate the pump foundation making the pump to develop vibrate, calling for stopping the pump and hence the production loss!

Even after the care taken to protect the structure; if it develops structural damage, it becomes mandatory to identify the root cause of damage identifying the chemical which caused the damage and then its structural rehabilitation method is to be designed and executed with that base. NDT of damaged structure helps in designing the structural rehabilitation scheme.

As a post treatment action, protecting the rehabilitated surface should not be missed out to ensure that rehabilitated structure performs its desired functional life.

Construction chemicals and structural rehabilitation.

Of course, structural rehabilitation and construction chemicals walk hand in hand and by matching shoulders to shoulders.

Rehabilitation of a damaged structure is a process like treating a wounded patient which needs some external medicines, some injectables and some finishing bandages. Similar is the process of rehabilitating a damaged structure.

Just as the treatment starts with laboratory tests for the patient, the tests (non destructive or semi destructive) are carried out on the structure to know the damage potential and severity. Immediately the role of (construction) chemicals begins.

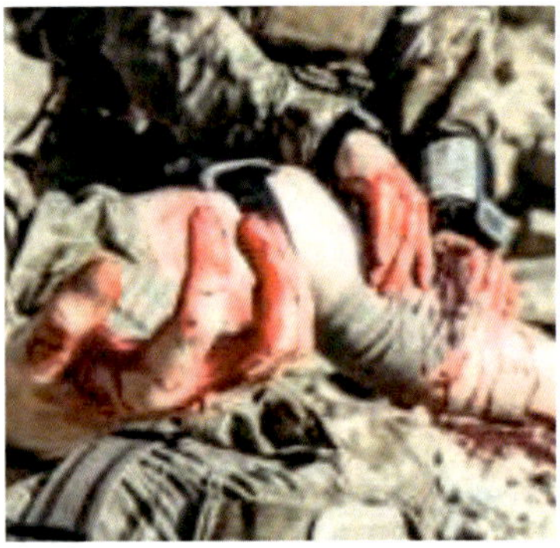

1. Wounds of corrosion on the concrete and steel structures are cleaned and applied with a proper coating for its healing and protection.
2. Some extra protective padding (similar to the cotton padding) in form of sacrificial anodes are then placed around the damaged portion of the structure).

3. Next stage of the treatment is injectables for the internal treatment which is for an invisible damage. This is of course with a purpose of increasing the immunity of the damaged member to resist further abuse.
4. Once all these actions are done, then comes a final stage of bandage. Just as it is done with the help of cotton and medicines in case of a wound, it is done with chemicals and additional reinforcement for damaged concrete members.
5. Curing the damaged human organ or a concrete member is the ultimate goalpost which is then achieved with the set of medicines / chemicals and instruments / tools and tackles.
6. Repairing a wound and rehabilitating a damaged concrete member have similar analogy which is incomplete without medicines / (construction) chemicals.

Construction chemicals and construction additives / supplements are often treated with similar thrust while both have different functioning. Construction chemicals have more effectiveness to "heal the wounds" of the structure while construction additives / supplements augment particular functionality or property or parameter for which it is used.

This makes a sense when construction chemicals are to be used for a chemical / petrochemical / agrochemical industry / refinery / dairies etc because compatibility of chemicals for their reactivity with the chemicals present around the damaged members stands on highest priority of all selection factors.

Consequences of delayed actions for structural rehabilitation.

Damaged concrete and steel structures are required to analyses, assesses, inspected and then to be essentially to be taken up for the appropriate measures for ensuring that their structural health is well maintained and they are functionally available throughout their desired functional life.

These actions can be of case specific nature like rehabilitation or retrofitting or refurbishment of strengthening as per the outcome of their damage assessment.

Once the damage is identified under the preliminary visual observation, it needs to be closely inspected so arrive at its seriousness and urgency to attend it. If it is identified to be normal and of non-structural nature, then cosmetic treatment may serve the purpose but if the damage is identified to be structural nature, then it is to systematically designed for upliftment of its functional life.

Delay in taking these measures (as designed) have serious consequences.

The damages like concrete corrosion which is de-polarisation of concrete matrix caused by many reasons; have tendency to grow and propagate to other parts of the member/s without showing any signs on the surface.

This is the most dangerous phenomena.
Such silent damages seldom give any sign of increase in damage till the structure fails or collapses. Time

duration between first large crack and final collapse at times, is not even few minutes. More often than not, such failures happen in concrete structures because of their plastic nature with limited elasticity. Concrete corrosion results in quick collapse of a structure after it crosses its plasticity limit.

Unlike this, steel structures have tendency to give indication of a major collapse as compared to concrete structure which never give such pre-indication with significant time lag. Majority of reasons of failure of steel structures is corrosion. This is visible on the surface. The structure with deteriorated structural parameters tends to bend and buckle at a slow speed which is captured under any visual inspection. It gives time to initiate and implement the rectification actions to prevent the collapse.

It is hence necessary, mandatory and advisable for one and all facility owners to ....

- Regularly inspect the structure and record visual observations.
- Catch any sign of uncharacteristic symptom on the structure and get it examined by the expert.
- Act as advised by the expert.

Dos and Don’ts for structural rehabilitation.

- ✓ Dos

1. Design the structural rehabilitation scheme and methodology after systematically assessing, evaluating and analysing the structural damages.
2. Ensure availability of total quantity of tools, tackles, equipment and construction consumables before starting the work.
3. Ensure overall safety measures for the structure when a part of it is being taken up for rehabilitation.
4. Ensure transfer of load of additional rehabilitation mass up to the foundation.
5. Ensure compatibility of construction chemicals with surrounding environment and core conditions.
6. Ensure appropriate and optimum pressure for injection grouting activity commensurate with the properties of material which is being injected.
7. Ensure scaffolding, strutting and supporting the formwork to receive free flow rehabilitation mass because its heat of exotherm exerts huge pressure on the formwork.
8. Ensure application of resinous applications only under moisture free conditions.
9. Protect the rehabilitated structure with protective coating to safeguard the rehabilitated structure from corrosive conditions.

× Don’ts:

1. Ensure that the rehabilitation activity doesn’t cause environmental hazards.
2. Avoid total thickness build up of more than 50 mm for application of polymer modified mortar.
3. Avoid loading the rehabilitated member before its stipulated setting and curing period.
4. Avoid execution under humid or high temperature period.
5. Avoid over-safe design for structural rehabilitation as it is uneconomical and may prove to be unsafe for the rehabilitated member at long run.

Design engineering and applied knowledge – two essentials for structural rehabilitation.

Whenever concrete or steel structure is found damaged, its damage assessment followed by analysis / evaluation of the assessment is the first step done.

Then methodology for retrofitting / rehabilitating / strengthening of the member/s is prepared.

This requires a fundamental knowledge of design engineering for quantifying the quantum deterioration of respective parameters of the member.

Execution of the assignment with designed method, requires more of an applied knowledge. Basic steps to be followed for this execution include the following:

1. **Surface preparation,**
   Experience for assessing the force required to prepare the surface by scabbling or chipping off the damaged surface; is the key for this step of work. This is achieved with having gained applied knowledge.
2. **Pre-execution action,**
   Prepared surface would require priming as per design requirement. Preparing the priming material and its application is better judged with experience. This is achieved with having gained applied knowledge.
3. **Execution of retrofit / rehabilitation / strengthening activity,**
   This is the most important stage. The properties of material selected for retrofit / rehabilitation / strengthening activity play a pivotal role in its

preparation for application. Factors like viscosity, yield, honeycomb combat, handling exotherm during hydration etc are few of many factors which can be better controlled under close visual observation. This is achieved with having gained applied knowledge.

4. **Post execution care.**

   Any structural rehabilitation task is incomplete without post execution care. Exotherm dissipation, curing of rehabilitated member, making the member ready for reinstatement of final finishes etc are the factors covered under post execution care. As like factors mentioned above, even this is not a part of any text book. This is achieved with having gained applied knowledge.

It is hence mandatory to amalgamate design engineering and applied knowledge in appropriate proportion for a successfully carrying out the task of retrofitting / rehabilitating / strengthening the damaged concrete / steel member in better than the best manner.

Importance of structural design dynamics in rehabilitation of damaged structures.

Damaged structures are first examined for their rehabilitation rather than their demolition and reconstruction.

Designing the scheme or method for structural rehabilitation of such structures demands a special address for structural design dynamics.

Parameters of the damaged structure often reach close to their reject limit and to bring them well within their acceptable limits, demands augmentation of those parameters. This is usually done with external treatment but this is to be validated with the structural design.

For example, if the structure is identified to be weak in shear at a particular location, then its structural design mechanics can be analysed to find out effects of this weakness on other parameters like bending moment or split shear or ductility etc. This is essential because while designing augmentation measures for weak parameter of shear, it should not happen that other parameter is ignored or not considered for its healthiness. It is imperative that while designing the rehabilitation scheme for compensating one parameter, some other parameter is adversely affected to create new area of structural weakness.

Structural rehabilitation is a 360^0 process which requires all round analytical approach to design a sound scheme for making it totally successful when implemented at site.

Structural dynamics and structural rehabilitation

Structural dynamics is fairly controlled by the strategic design parameters.

When a structure undergoes corrosion, - leading to structural damage, the restoration / retrofit / refurbishment process is initiated.

Structural dynamics control or guide and channelize a method for rehabilitating the damaged structure.

The parameters like bending moment, split shear, punching shear, uplift pressure, differential lateral thrust etc are few of the many conditions which can damage the structure in addition to corrosion.

The speciality of structural damages due to the disorder in structural dynamics; is that they don't give an apparent sign of structural distress. The signal through which such distress is identified is deflection or corner cracking etc.

While designing the structural restoration / retrofit /rehabilitation method or scheme for such structures, adequate importance is to be given to structural design and dynamics. It doesn't just end with a conventional method of removal of some deteriorated mass, some injection and some reinstatement.

It calls for addressing the disturbed load transfer phenomena, unconventional behaviour under software analysis while subjecting it to anticipated loading condition, safety factor etc.

A small but significant part is also extracted at times from special properties of construction chemicals. For example, injection of glass flake filled low viscosity high molecular weight epoxy into concrete foundation for high RPM machine can grossly mitigate its vibration which – if not controlled; - can structurally distress the concrete foundation and might result in stoppage of machine operation.

A call is required to be taken by the construction chemical manufacturers to come up with a range of suitable construction chemical/s which can be successfully used for compensating the distress caused by disorder conditions like punching shear in tension zone of concrete.

Importance pressurized operations for grouting nature during rehabilitation of damaged concrete structures.

Structural rehabilitation of damaged concrete structures often calls for injection of specified grouting material inside concrete core. This injection is always done with specified pressure. How important that pressure is?

It is indeed very important.

When an under-pressure activity is to be taken up, its necessity is to be properly addressed. Injection method is to be sought for only under special situations where other methods are not found effective. Injection operation under pressure is a blind work. One won't know the path followed by the material being injected although it would have a reason for its selection based on observations and results of non-destructive testing. This calls for its meticulous execution at field so that the activity happening beyond our vision, happens to the benefit of the damaged structure.

Pressure at which the grout is to be injected within the core concrete must be very precise and optimum. It should be just adequate for the grout (with its known viscosity) to reach and to plug the crevice of the core concrete mass. It should also be not very high for pushing the grout for plugging the crevice because if it is higher than required, it would exert additional pressure on the walls of crevices as a result of which the crevices would expand before getting plugged. Of course, the crevices would be plugged during the injection process but (as said earlier,) it may create the stresses due to excessive pressure in that case through the crevices which might add a new dimension to the damage.

Always be careful while carrying out injection operation to rehabilitate the damaged concrete members.

Insufficient curing for rehabilitated concrete members.

Rehabilitation / retrofitting / strengthening of concrete members is carried out with help of high-grade materials. All of them have tendency of high exotherm generating during polymerization of their ingredients. This has to be brought to balanced thermal equilibrium. Dampening of the surface is the usual method adopted for the purpose.

Curing activity is also called as strength gaining phenomena.

Two points of importance connected with curing are discussed here.

First point is insufficient efforts to compensate for polymerization exotherm. This results in generation of internal and surface cracks through which exotherm heat tries its escape way. Unfortunately, some of these cracks turn to be structural cracks at long run and emerges as serious structural defect.

Second point is insufficient time period provided for full polymerization of total quantum of ingredients of the rehabilitation matrix which is also covered under the domain of curing. Polymerization period for each category of rehabilitation material is specific and defined. It is mandatory and essential to allow total polymerization time period. All designed and desired parameters of finally rehabilitated structure are achieved only after this polymerization process is complete. If under any circumstances this period is not allowed before putting the structure to its desired functioning, there are serious chances (with high probability) that the structure might buckle and no longer remain capable to sustain the load and other stresses to which the structure would be exposed on later date. This is very typical challenge. No NDT result would be below rejecting limit. No reinforcement would be found corroded. No external force causing distress would be identified and still the structure may show hollow sound at some local pocket! Internal distress due to buckling or typical shear is the reason for this and insufficient time allowed for polymerization process - curing period - is the reason for the distress as stated above.

Any compromise in curing activity and curing period is a serious threat to the rehabilitated structure which was already damaged and which has been rehabilitated.

Crying cracks say ….

We often find cracks – structural or non-structural - in constructed elements. Distinction between both the types need a thread bear Technical analysis.

Cracks occur because of these reasons.

1. Pressure from one side of the element which can't be resisted by the element as a whole.
2. Stresses generated within the element.
3. External thrust or impact,
4. Differential settlement / expansion / contraction (shrinkage).
5. Humidity ingress.
6. Change in service condition of the element.
7. Distress in under designed structures.

There could also be few more reasons for crack formation in the element.

Crack is a separation plane within a homogeneous element which is beginning of the distress within the element. While depth and width of any crack can't be ignored, its formation can't also be neglected because it gives an avenue or access for deteriorating elements (gaseous or liquid or solid) to enter the element and reach its core. If this element is hygroscopic in nature, it makes its approach easier to reach the core of concrete. Methods are available in the world of non-destructive and semi destructive testing for reaching the reason of generation of crack. Successful sealing of crack can be declared only when it doesn't reoccur.

Bridging the crack can be done in many ways but if it is done without knowing its reason, it might reoccur. It is noteworthy that more often than not just sealing the crack or bridging it with a filler paste or a shrinkage free compound may not be enough, because that is not a scientific repair of crack.

Performance of micro-concrete with addition of 5 mm down aggregates:

Micro-concrete is one of the most versatile materials for structural rehabilitation / retrofit activities. At times it is added with 5 mm down size aggregates for specified purpose/s. How does it perform?

A. Compressive strength:

It is a principle which is taught in classrooms of Civil Engineering education that compressive strength is directly proportional to maximum size of aggregates. Maximum size of aggregates in micro-concrete (without any addition) is 2 mm. If this maximum size is increased to 5 mm by adding such aggregates, the compressive strength of the matrix is bound to increase. Although it is not mentioned in any of the text books, it is experienced from the long field experience on the subject that compressive strength of micro-concrete added with 5 mm down size aggregates in 1:0.5 proportion (1 - micro-concrete and 0.5 - 5 mm down size aggregates) is 20% higher than compressive strength of matrix with only micro-concrete.

B. Exotherm /Heat of hydration.

It is a principle which is taught in classrooms of physics that thermal energy dissipates with increase in area of its domain. When 5 mm down size aggregates are added to the matrix of micro-concrete, the exotherm / heat of hydration generated is dissipated in larger specific surface and hence its effects of expansion causing thrust / pressure on shuttering get proportionately compensated. Although it is not mentioned in any of the text books, it is experienced from the long field experience on the subject that exotherm / heat of hydration of micro-concrete added with 5 mm down size aggregates in 1:0.5 proportion (1 – micro-concrete and 0.5 – 5 mm down size aggregates) is 07% lower than the exotherm / heat of hydration generated in matrix with only micro-concrete.

C. Limitation.

Additional of 5 mm down size aggregates to matrix of micro-concrete has one significant limitation. The matrix with addition of 5 mm down size aggregates starts showing micro porosity. This is logical. Specific surface of micro-concrete then has to cover large size particles for which the ingredients of micro-concrete may fall short.

D. At the end:

Addition of 5 mm down size aggregates to micro-concrete has some distinct advantages but also has specific limitation. It is recommended to add 5 mm down size aggregates to micro-concrete in designed proportion (1:0.5 or 1:1 or otherwise) where incremental compressive strength and significant reduction in heat of hydration is recommended but it is to be critically reviewed when it is to be utilised for rehabilitation of the concrete members which need near to zero porosity. This is a basic requirement for a chemical industry where micro porosity can allow corrosive gases or corrosive liquid to enter the core concrete through micro porosity due to addition of 5 mm down size aggregates.

Micro-concrete is one of the versatile materials used for structural rehabilitation activities. It has many important properties like free and convenient flow within formwork, early strength gain, reduced water permeability etc.

At times aggregates from 5 to 10 mm size are added to the matrix of micro-concrete. The reason for adding aggregates must be clearly understood before taking this step. The subject is briefly discussed below.

1. For most of the times, addition of aggregates helps for economising the cost of micro-concrete - being an expensive option as compared with conventional concrete.
2. Addition of aggregates increases the strength of micro-concrete matrix so when it is desired the strength of micro-concrete matrix (after it completely cures), needs to be higher than as specified by the manufacturer, aggregates to the maximum of 100% (1:1 proportion) are mixed with micro-concrete.
3. Addition of aggregates has significant effect on water absorption of completely cured micro-concrete matrix. It increases the porosity of the matrix by some percentage which is a significant point of review when the rehabilitated member is going to be exposed to corrosive conditions (saline / chemical or other such conditions).
4. Quality of aggregates require close inspection to ascertain that they are non deleterious and free from impurities, Non compliance of these parameters for the aggregates mixed with micro-concrete can affect the ultimate desired performance of micro-concrete.

Structural rehabilitation, stringent timeframe and facility downtime

When any structure is taken up for its strengthening or rehabilitation or refurbishment or retrofitting for its rectification, the point of nonavailability of the structure or facility stands on high priority. This is also called as outage of the facility or the structure. Although it is more significant for industrial sector, it is equally important for residential or commercial or other facility or structure.

Planning of structural uplift action for any facility must be invariably made considering the time for which it won't be available for use. While this outage of the facility calls for optional arrangement for housing the functional activities happening in it, expenses would also include shifting and re-shifting the infrastructure, effects on output of the occupants etc. When it is considered for any industry, it directly affects the production loss and hence more often than not such activities are differed unless they become utmost essential bringing the structure to the threshold of 'unsafe' category.

Whatever the structure may be, it is essential to have a strict control over the execution schedule of structural strengthening or rehabilitation or refurbishment or retrofitting such that down time or outage of the facility housing the activity would not be prohibitive for it.

Timely execution of any retrofit activity is a service to Nation. It is economical, it saves energy and resources, it reduces outage of availability of the facility, it gives an opportunity to take up further more assignments, it makes the client of the facility owner satisfaction and happiness.

Importance of coving at junction of vertical and horizontal member.

Junction of horizontal plant (floor in a structure) and vertical surface (wall of a house) is often found passing the liquid which comes close to it. The passage may either be from positive side or from negative side but the junction is clearly found allowing passage of liquid.

A simple step can be helpful in preventing this phenomenon which has always remained an eyesore.

Two planes meeting each other at right angle generate a stress point. It becomes a weak spot and it can't sustain any stressful condition from any of its sides. Reason of leakages through basement floors or ground level floors which are not elevated with plinth fill; often have this problem. Junction of terrace floor and parapet wall significantly faces this problem.

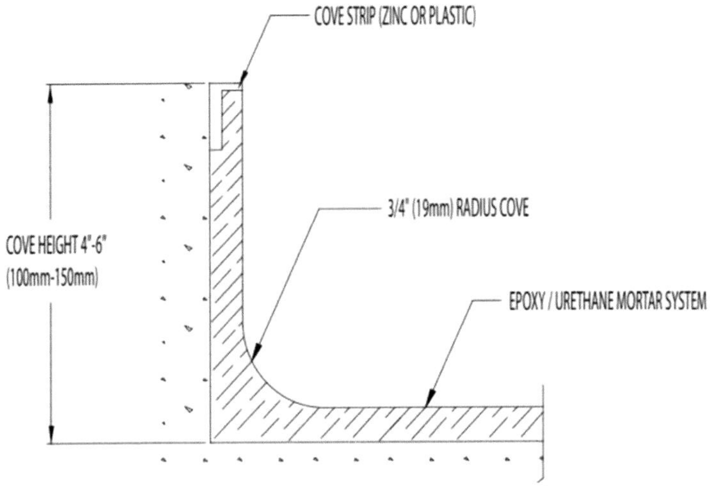

If the junction is provided with a fillet type of arrangement (which is also known as coving and is the technique of installing small barriers lining the edges of a floor where it meets the wall (referred to as coves); the stressful spot will not occur. Coves are generally rounded 45-degree angles which slope upwards, preventing a hard 90-degree angle where floor and wall meet. Material of coving may not necessarily be epoxy or PU (as shown in figure) but its top must be behind external face of vertical plaster and its horizontal end must be above the floor so that water flowing from vertical floor to horizontal floor through coving, passes without any obstruction and penetration through the junction of vertical and horizontal plane.

Damages to elements (human body part and Reinforced Concrete) and its treatment:

1. Reinforced Concrete consists of aggregates bound under cement paste and reinforcement steel.

2. Cement paste is a result of physical reaction followed by a chemical reaction.

3. Bond of reinforcement with concrete matrix; makes it an integral part of reinforced concrete.

4. Damage to the reinforced concrete matrix is similar to a wound on body. It may either begin from ingredients of concrete or from steel but once it begins, it spreads like cancer and covers the complete member.

5. Treatment of damaged concrete begins with reaching at the root cause of the damage for which testing is necessary just like pathology laboratory tests for human body. This is essential to ensure that concrete matrix wound won't repeat with the same root cause.

6. Little differentfrom human anatomy, treatment of damaged concrete can't replace the damaged part in totality. Hence its core has to be strengthened or treated for the identified damages and certainly its skin is to be repaired and strengthened.

7. Skin concrete (cover concrete) retrofit has to be designed and executed with a very special concept. **Skin concrete (cover concrete) encompasses or envelops the core concrete.** This very phenomena play very vital part in rehabilitating the damaged concrete member. Even if it is not feasible to rectify few inferior parameters of the core concrete, the extra strong skin concrete (cover concrete) will encompass / envelop the core concrete and the member would be available through out its designed functional life.

Protecting concrete and steel from corrosion.

As we all need protection from climate changes like Woolen ware in winter and raincoat in monsoon, concrete and steel structures also need similar protection to resist the challenging conditions.

Challenges for concrete needing protection are like this.

1. High humidity,
2. High temperature,
3. Saline conditions,
4. Reinforcement corrosion,
5. Design inadequacy,

Challenges for steel needing protection are like this.

A. Corrosive fumes or splash,
B. Uncalled for stresses generated within the structural matrix,
C. High humidity and temperature,
D. Modification in functionality of the structure.

Both concrete and steel structures are prone to corrosion. Corrosion of concrete is defined as depolarization of cementations or binding property. Corrosion of steel structure is occurrence of rust on its surface.

Concrete structures:

They are exposed to the challenging conditions at regular intervals. If UAE perspective is considered, then thermal and humidity variations are significant in this region. Hence the methodology for protecting them against getting corroded can be largely divided into two major categories.

In-build resistance to corrosive conditions,

External measures.

In-build resistance for concrete matrix can be created by following measures.

i. Construction by following standard construction practices,
ii. Treatment of construction joints and expansion joints with utmost care.
iii. Structures with special services like water retaining structures, retaining walls, foundations for heavy machinery, mass concrete etc must be constructed by following guidelines provided in Global codes of practices.
iv. Use of construction chemicals should be very critically evaluated before taking into execution.
v. Functionality of structure must match the structural design and execution.

Reinforcement corrosion happens for RCC members in corrosive surroundings. Lot has been discussed on this subject. An effort is being made here to add some new dimension to the subject.

Increase in volume of steel is one of the most significant effects of reinforcement corrosion. The incremental volume exerts pressure on compact core of concrete which gives rise to crack in concrete member and that is beginning of the end.

It is necessary to understand the factors which have caused corrosion. Corrosion caused by chemical conditions like penetration of corrosive chemicals within concrete mass, salinity, humidity etc. possess potential to corrode reinforcement steel of RCC member.

Corrosion caused by adverse chemical conditions has to be examined under close inspection. This is because if corrosive chemicals or conditions have been able to affect reinforcement steel, they would also have a potential to affect concrete mass and corrode concrete too. This corrosion affecting concrete can be noticed only after concrete is fairly damaged. Extension of reinforcement corrosion to concrete is a process of interlink of potential hazard of concrete deterioration. Rust which is trying to get accommodated ends up in cracking the concrete matrix. The space created by this cracking then stores the corrosive elements. Stagnancy of these corrosive elements within concrete mass gives them an ample opportunity to react with concrete ingredients and deteriorate the concrete matrix.

The process is blind hence it can be noticed only after a big damage is done and failure is just on its threshold. The root cause of such mysterious damage to concrete member is reinforcement corrosion. Hence let us be very cautious in mitigating reinforcement corrosion.

Commercial perspective for Structural Retrofit.

Whenever a structure undergoes its rehabilitation or its retrofit, it progresses to achieve additional functional life. Its survival period is increased with this action.

As is evident, every action has a cost and effort. This is also true for the action of rehabilitation or retrofit action for any structure.

It requires a review for this aspect to begin with creation of the structure. A specific effort and cost has already been made in creating the structure. Necessary, required and adequate resources have already been utilized to create the structure in addition tot eh efforts of Engineering, nurturing, reviewing and perfectionating the conceptual stage for the structure. All these actions also might have attracted a commercial input. In short, the creation of any structure happens with the multi commercial aspect which covers the objects in front and objects behind the scene.

If the structure deteriorates due to any reason - more often than not - it is taken for its rehabilitation or retrofit action. The methodology for this action finally undergoes the scrutiny by a commercial team and a justification drive for the cost at which the action is estimated is reviewed with necessary criticality.

While cost optimization is an essential exercise for any such action, it becomes furthermore essential point of view to strike a fair balance in the exercise. The cost at which the rehabilitation action is conceptualized, is inclusive of all the interlinked activities required for ultimately bringing new life to the structure. If any of these activities is targeted during this cost optimization drive for questioning its necessity of functional need, the thought process starts tending towards its elimination from the entire scheme prepared for the rehabilitation action. Commercial considerations might – in the process – miss out an important link for the ultimate success of the effort.

Another side effect of cost optimization brings commercial negotiation in the picture. The creator of the concept is asked to reduce the individual item cost or overall cost which is estimated for the exercise. A hard pressed efforts under this commercial negotiation might fetch some cost reduction till the negotiator gets satisfied.

How does it influence the holistic approach for this activity?

When the rehabilitation or retrofit action is conceived with a set of materials and methodology for their application, the new commercial format, forces the executor to either replace the material from the material with which the action was designed or create a team of skilled and semi-skilled workers with a step lower efficiency to suit to the revised budget for the work or to reduce the profit margins.

None of these end effects of hard commercial negotiation favor the execution of this action with the desired quality. Second grade material won't give the desired results. Second grade set of execution team of skilled or semi-skilled workers are prone to commit mistakes in sequence or other aspects of adherence to the desired quality for the work. Reduced profit margins would certainly reduce the interest of the executor resulting in the reduced overall attention of senior management of the Organization which is the most detrimental thing for the most important action of rehabilitating the damaged structure.

Having said this, it is not the intention to carry out the activity with an exorbitant cost. The action of cost optimization is necessary even for this important action, but it should be done with an overall holistic view. Few considerations which might make the commercial negotiation for this action more effective are briefly addressed below.

- Hard negotiation should be avoided. The creator of the concept of rehabilitating the structure has already surveyed the structural health parameters of the deteriorated structure and hence he / she is aware of the exact treatment to be given to the structure for enhancing its functional life. Hence putting questions on necessity of the items and their costing might not be necessary.
- Instead of putting stress on cost of the work, the stress should be applied to the important factors for the assignment. Few of them might be critical for their timeframe for execution and few might require ultra-high quality for execution.
- The assignments with stringent time frame may be linked with a strict "bonus penalty" clause which demands the focused attention of the executor to plan and execute the actions within

the stipulated time-frame. This needs a set of extra efforts which ultimately boils down to commercial input - a cost optimization or real value for the money spent.

- The structures demanding ultra-high focus on the quality are those which have lost their structural parameters much below their acceptable limits. This can be ascertained with the help of post rehabilitation non-destructive tests (NDT) on the rehabilitated structures which drives the executor to put hard efforts on quality assurance aspects which again attracts the appropriate commercial input – another face of a cost optimization or real value for the money spent.

There might be few more points on which effective cost optimization can be achieved even without hard commercial negotiation.

What is the "take away" in the process of structured and balanced commercial negotiation?

1. Adherence of the necessary parameters affecting the assignment like time-frame, quality, downtime reduction etc.
2. Getting the value for each penny spent for the assignment.
3. Minimizing the opportunities of repetition of the work due to poor quality or left out work due to time shortage etc.

One important point, which needs to be addressed before ending the discussion on the subject. The rehabilitation action of any structure is a process of giving new life to the same which was on the threshold of "going to ashes". It is a systematic process in which the damages are scientifically identified, analyzed and evaluated which is followed by design of the most appropriate rehabilitation scheme covering selection of materials and their application methodology and post rehabilitation protective treatment which is to be religiously implemented.

The most important benefit under this exercise is that the cost of reconstruction of the structure gets postponed, depletion of resources in reconstruction action would be avoided, downtime for operating units causing production loss can also be avoided, the process can accommodate additional strengthening measures of the existing deteriorated structure which otherwise would have attracted additional Techno Commercial efforts, avoiding the cost of demolition and causing environmental turbulence etc.. Estimated cost for the activity should also consider the cost of hidden saving which is achieved with these benefits. They justify marginally higher unit cost of rehabilitation as compared with eh unit cost of reconstruction of the structure

Commercial perspective for any structural rehabilitation or retrofit assignment is to be designed with a professionally matured and fair deal of balance between Technical and Commercial disciplines because both of them are like two wheels of the vehicle called structural integrity.

Bond Action between mother surface and new matrix

Rehabilitation of any damaged structure is a process of strengthening its deteriorated parameters and lost structural properties like strength, integrity, density, chemical configuration etc. Difference between reconstruction and rehabilitation of a damaged structure is that reconstruction of a damaged structure requires its demolition while in rehabilitation, that is avoided.

The process of rehabilitation encompasses the fundamental concept of not only replacing the damaged portion of the member but also replacement of damaged parameters of the member. Reinstatement of fresh matrix over the damaged member to replace the damaged portion is one of the activities of prime importance.

Any reinstatement calls for a strong and tight bond between mother surface and fresh matrix being applied as a reinstatement measure. The tightness of this bond directly relates to the integrity of the rehabilitated member and its process. Establishment of this bond largely depends on these factors.

- Density of mother surface and that of fresh matrix should be matching or close to each other. Fresh matrix with high density might become an additional burden on the member under rehabilitation which is already damaged.

- Thermal coefficient of expansion and contraction of the fresh matrix and that of mother member should also be fairly close to each other. Although the mother member has been existing for some time and its thermal coefficient of expansion and contraction might have been stabilized, if the thermal coefficient of expansion and contraction of fresh matrix is at variance from that of the mother member, it might exert and additional stress on the mother member which is not usually considered during engineering design of the rehabilitation action.
- Self-weight of fresh matrix is extremely important because till the time of polymerization or hydration of the fresh matrix its load is to be supported or shared by the mother member which is weak and under rehabilitation.

Once the fresh matrix is selected based on NDT results and the criteria as stated above, its integration with the mother member is to be designed and to be accordingly executed.

Usually the fresh matrix consists of polymer modified mortar or self-levelling non-shrink micro-concrete.

Polymer modified mortar has less density than micro-concrete with self-weight also being less than that of micro-concrete. Hence efforts required to hold the polymer modified mortar till the time that its attains self-stable state are moderate and efforts required to hold the micro-concrete till the time that its attains self-stable state are required to be stubborn

Normally bond action in structural rehabilitation activity is provided by two processes. The first is bond coat of compatible material and another is a mechanical bond which is provided with structural arrangement.

In case of polymer modified mortar, being of limited self-weight, it requires moderate efforts are required to hold the matrix till it attains stable state. Resinous bond coat or polymerized cementitious bond coat is sufficient for holding mm thick layer of polymer modified mortar (maximum allowable thickness for each layer of polymer modified mortar is mm beyond which it tends to distract from the mother surface due to its excess self-weight. The process of applying bond coat for rehabilitation treatment based on polymer modified mortar is briefly mentioned below.

- Clean the surface to remove all loose dust, oil, grease and such contaminants which can obstruct development of bond between mother surface and the coating material.
- Ensure that the mother member is free of humidity and moisture right till its core by plastic paper test.
- Mix both the parts of resinous material in a container and continue to rinse / mince it well till uniform texture appearance of the resin is achieved.
- Setting time for this is very important. It is mandatory that the polymer modified mortar which is to be used for reinstatement is to be essentially applied when the bond coat is still tacky. This is the stage when it has started polymerizing and still under the process of

polymerization. The logic of the process is that polymerization of bond coat covers or influences the reinstated material (polymer modified mortar) to make it it's integral part such that the mother member and reinstated matrix are tightly bonded to each other and later on it would become an inseparable part of the mother member.

- ❖ All subsequent activities are to be carried out as designed or as per manufacturer's recommendation.

Another tool for bonding the reinstated matrix with the mother member is shear connector or shear dowel. Set of shear connector or shear dowel is used when the self-weight of reinstatement matrix is higher than that of polymer modified mortar.

It won't be possible for resinous bond coat with limited bond strength to hold the volume of reinstatement matrix with higher self-weight for the time till it attains self-stable state. Hence a mechanical bond is to be provided in this case. The procedure of this arrangement is placed below.

- ❖ Drill hole of sufficient diameter to place the shear connector / shear dowel inside the mother member. e.g. if the shear connector / shear dowel is of 12 mm diameter, the hole diameter should be 14 mm.
- ❖ Clean the hole to remove loose dust and other contaminants from inside.
- ❖ Place the L shaped shear connector / shear dowel inside the hole such that longer side stays

inside the hole and shorter side stays outside the hole.

- ❖ Fix the shear connector / shear dowel inside the hole with non-shrink polymer grout by hairdressing the grout inside the hole.
- ❖ While only the shear connectors / shear dowels may not be adequate to hold the fresh matrix of reinstatement, additional reinforcement as per design is to be provided which would be tied up with the end of shorter side of the shear connector / shear dowel.

It is necessary to select the most appropriate bond arrangement for the process of structural rehabilitation action. Few important points which should be strictly followed is placed below.

- ✓ The reinstatement matrix is to be applied over the member under rehabilitation within the settling period of the resinous bond coat when the bond coat is still tacky.
- ✓ If bond coat sets before placing the reinstatement matrix, it acts as de-bond coat and deprives the bond action between old and new surfaces.
- ✓ The shear connector / shear dowel should be placed at predesigned spacing to optimize their requirement based on the load to be supported by them.

Bonding the fresh matrix with the mother member is one of the most important activities for success of the action. It must be executed necessary skill.

Steel structures – Section 1 - Construction perspective:

Steel structures are constructed with specific functional need.

- They have better elasticity and ductility as compared to concrete structures,
- Their construction including fabrication and erection are faster than concrete structures,
- Their members can be taken for multiple use especially in industrial structures,
- Their load carrying capacity is higher than that of concrete structures except extra ordinary high strength concrete members,
- Overall economics for steel structures as compared to concrete structures could be favourable for heavy structures etc.
- There could be still more advantages of steel structure over concrete structures.

With all these advantages, there are few basic points of care to be essentially taken for fabrication and erection of steel structures. They are as mentioned below.

➢ Rerolled steel sections have inferior metallic parameters as compared to fresh steel. Hence if it is not possible to avoid their use, care of specific inferior parameter should be taken at design stage.

➢ Steel sections lying in open yard (exposed to heat or rain or saline atmosphere) would get corroded with passage of time hence they should be checked for effects of corrosion on their body before taking them for use. It is likely that section thickness (of web or flange) might have reduced due to corrosion which needs to be appropriately addressed.

- Welding process and post welding care are well defined in respective International standards so they have to be holistically followed.
- Steel structures have higher coefficient of thermal expansion and contraction than concrete members. Accordingly, movement happens in a steel structure with thermal variations. It is hence extremely important to allow this movement at one end of the fabricated member – especially horizontal member and truss. Method to accommodate this movement is to provide bolted connections at one end and welded connections at another end if welded connections are necessary from rigidity point of view (structural design requirement). Absence of bolted connections have at times resulted in chaotic situation for few structures.
- Condition monitoring of steel structures from corrosion assessment point of view is a prime point of post erection care. Steel structure has to be applied with anti-corrosive painting with suitable painting system to protect them against corrosion caused by chemical conditions, UV rays, atmospheric salinity, thermal variations etc.

STEEL STRUCTURES ARE INDEED A BOON FOR STRUCTURAL AS WELL AS AESTHETIC DESIGN OF MODERN TIME.

THEY MUST BE WELL PREPARED AND MAINTAINED.

Steel structures – Section 2 - Rehabilitation of steel structures:

Steel structures are more prone to corrosion under corrosive conditions as compared with concrete structures. Rate of deterioration in form of loss of metal volume (thickness of member) due to corrosion is distinct for steel members whereas loss of parameters of concrete members due to corrosion, requires analysis and investigation for quantifying them.

With loss of volume / mass of steel in the member, it will reduce load carrying capacity of that member and that makes the member unsafe from were rehabilitating that member, becomes necessary.

More often than not it is observed in chemical, petrochemical, fertilizer, agrochemical industries and refineries that such members whose steel volume has reduced beyond certain limit are discarded and replaced by new members. This is a time-consuming process and also invites loss of production which makes it further uneconomical.

Steel members can survive till loss of their parameters in higher limits as compared with loss of parameters of concrete members; which gives a transition slot for corroded steel members to analyse them and examine them for their rehabilitation.

Corroded steel structures are to be thoroughly examined for loss of their parameters and their resultant effects in supporting type of stresses (tensile stress or shear stress or flexural stress or bending moment etc). These stresses are maximum in magnitude at certain distance from end of the member. Hence loss of parameter at certain distance from end of the member, holds a key to get an idea for possibilities of rehabilitating the member. If loss of integrity for the member is within its acceptable limit, then augmenting steel volume / mass at such strategic locations (as explained above) can be a bright option. Accordingly, the data of lost parameters can be fed to the respective software for calculating the volume / mass of steel to be augmented and the location at which its addition would be meaningful.

This is indeed an exercise which demands very focussed approach. When economics of replacing the steel member v/s rehabilitating it are compared, this exercise is very useful.

Technicalities can't be sacrificed for better economics but if the Technicalities can be optimised with use of applied knowledge with passion to save the precious natural resource like steel, it would be a real patriotism for an Engineer.

Steel structures – Section 3 - Anti-corrosive protection measures:

Steel structures have very specified use for industries and domestic sector. In addition to their advantages and merits, they also have few limitations. One of the major limitations of steel structures which at times make it prohibitive for their selection; is its easy susceptibility to corrosion under conducive conditions.

There are means and measures to protect them from corrosion. The most common of them is anti-corrosive protective coating / painting.

It must be understood very well that cosmetic / decorative painting is different from protective anti-corrosive protective coating / painting.

Steps to be holistically and strictly followed while carrying out the activity of anti-corrosive protective coating / painting are briefly placed below.

1. Anti-corrosive protective coating / painting should be carried out under controlled humidity and atmospheric temperature. High humidity and high atmospheric temperature affect polymerization of ingredients of paint matrix and the finally desired end properties are not achieved.
2. Surface preparation contributes 80% of total quality of the entire activity. Whatsoever grade of painting system might have been designed, if it is applied on under prepared surface or humid surface, it won't perform as desired till the designed life.

3. The mixed paint material must be well stirred before every pass of brush / roller. This is with a specific reason that solid content of the paint matrix which is supposed to be applied on the steel surface with liquid career (which is solvent and which evaporates with time) must be completely applied on the steel surface and should not be left within the paint container - making it heavy weight piece (due to settled solid content of the paint matrix - which should have gone to the steel surface but has remained in the container).
4. It is recommended to apply two coats such that the subsequent coat is perpendicular to the previous one.
5. Brush or roller must be hard pressed during paint application against the steel surface so that effective adhesion of paint material is established right from the time of application.
6. Mixing ratio of two or more components of the paint material must be very strictly followed as prescribed in the Technical Data Sheet of the material. Deviation from the prescribed mixing ratio leads to improper polymerization process which results in improper cross linked (paint) film formation.
7. Desired DFT and WFT must be achieved during the paint application activity.
8. Painting system with higher WFT should never be thinned with the thinner to trim down the DFT of paint film. It would rupture the paint matrix during polymerization.
9. The painted surface must be allowed to cure and polymerize before putting it to use / expose to the corrosive conditions.

10. Heat resistant paints cure / polymerize only on attaining the range of temperature on the steel surface. Pre-application of heat resisting paint and expecting it to perform its function of heat resistance won't fetch desired result.

Safety significance:

Safety is one of the prime essentialities for structural retrofit activities. This is because they are to be carried out on and around those structures which are already distressed. Factors due to which the safety is threatened, broadly cover deteriorated parameters of concrete and steel.

Mitigating unsafe conditions is also a type of safety

adherence. Structures of UAE have emerged with typical unsafe situations which have specific seriousness.

Water ingress in concrete members is one of the most common defects in structures. Seriousness of this defect attains more gravity because of quality of water

entering the concrete matrix. Probabilities of water which has easy access to core concrete is contaminated. Concrete corrosion is then followed by reinforcement steel corrosion.

Reinforcement corrosion is a grave threat to the structure.

Mitigating water ingress in concrete members is safety compliance for the structure. Protection of the structures with compatible coating and water-resistant treatment helps in this. External and internal waterproofing for basement with a tailor-made design for each structure is the appropriate solution for this purpose.

Another safety related threat for structures is slenderness ratio. Limitation of land space at UAE has driven the owners and consultants to actively precipitate the option of high-rise structures. While utmost care is taken during structural design of such malty story building, limitations of space, architectural / aesthetic values etc influence on the dimensions of the building. This can be bringing the structure to unsafe zone. Structural safety is required to be considered a step ahead of cosmetic aspects of the building.

MEP services in the buildings have often been a reason of leakage of water and electrical power. If sufficient time is not allotted to the activity, joinery in this service suffers some compromise. It is a requirement for a safe structure that all of its MEP services are perfectly laid such that they don't leak throughout their designed functional life at par with the life of the structure.

Structural rehabilitation and structural safety

When a damaged structure is taken up for its retrofitting or rehabilitation action, the address is to bring it to its normal parameters with optimum cost and effort. More often than not the structure is retrofitted in segments or parts. The most affected area is considered as benchmark and surrounding areas are provided with the befitting treatment.

Narrowing down the width of the subject, when a scenario develops when only 3 out of 7 columns and 2 out of 5 beams are to be attended for retrofitting / rehabilitation action, the delicacy or intricacy of the matter attains a close focus and sharp watch especially on the structural safety.

Rehabilitating any structure is part or segment transfers stresses to the members which are not covered under this exercise. Stresses often are allowed to square up at the members which are not a part of rehabilitation action. What happens is that these members are not provided with any additional supporting treatment and suddenly they are exposed to the stresses generating due to damaged members under their rehabilitation. Load share by the members under rehabilitation reduces and gets transferred to the members out of the treatment. This is indeed a safety matter for the unit as a whole which can be called as structural safety.

Matter attracts higher concern when partly rehabilitated structure is put to service and load sharing as per original design. Rehabilitated members might have better parameters and members which were not a part of this exercise might have the parameters which might be at variance from the improved parameters and if that happens, it is a serious structural safety concern.

Structural rehabilitation and safety are directly related with each other. Usually, the structures which are pushed to 'unsafe' zone are taken up for their rehabilitation so safety stands on highest priority for all activities under structural rehabilitation.

1. **Design Stage:**
   While designing the rehabilitation method / scheme; cognisance of test results (destructive and non-destructive) is essentially taken. The parameters which are deteriorated are identified under the exercise and the design is prepared to recoup those lost parameters. Care is taken for designing the rehabilitation methodology for critical structures like industrial structures, structures in high-risk seismic zone, coastal structures etc to provide an incremental factor of safety. This is to sustain of higher magnitude of damage expected to affect the structure.
2. **Pre-requisites stage:**
   Once the design is ready, specific and tailor-made pre-requisites are implemented at site. This is the stage where maximum thrust is given to safety measures. If only part structure is to be taken up for rehabilitation, then safety of other part is of high importance to ensure that its stability and soundness is not endangered during rehabilitation of the damaged part of the structure. The part of structure which is not to be taken for rehabilitation is to be properly propped and tied up with external means such that it does not develop structural damages.

3. **Execution stage:**

The most important stage is execution stage. Removing the damaged portion from the member should be carried out to NOT to exceed standard provisions. Time of the total activity of rehabilitation forms a governing factor in deciding the safety factor to be attached to it. Monitoring the pressure for injection grouting, erecting fresh reinforcement, managing heat of hydration / polymerisation of ingredients of rehabilitation matrix, ensuring strong and sound formwork with strutting, (compensating the pressure exerted by rehabilitation matrix on the

formwork tends to deform the formwork) etc are to be very meticulously executed by reviewing each one through critical eyes.

4. **Post execution stage:**

Safety precautions don't end with completion of the execution. Post execution care is equally necessary to end up the entire activity without a safety mishap. De-shuttering attracts maximum attention. Reinstatement of finishes or services on rehabilitated surface requires specific address to safety factor to ensure that it is carried out after the member has been completely cured and hydrated.

Construction Management

Managing construction of any magnitude needs a focussed attention for completing the assignment within time frame and budget with all necessary safety measures.

Construction Management has many tools and faces which are ultimately to be interlinked to make the assignment successful. Few of them are briefly discussed below.

1. **Safe and conceptually clear structural design.**
   The structural design of the unit to be taken up for construction has to be very clear and safe. Any ambiguity or inadequacy or incomplete detailing leads to loss of time and efforts for necessary rectification. The pre-requisite for safe and clear structural design is conceptual clarity with the facility owner. Functionality of the unit to be taken up for construction must be well defined and conveyed to the structural designer so that the factors to be considered for structural design are properly addressed and nothing is left unattended. Change in functionality than the one which was conceived will have its effects on structural design. A commercial complex can't be used as an industrial unit.

2. **Resource Management.**
   Once the structural design and drawings are available for execution, then the next step Is to identify the resources for execution.

Resources compile of two types – skilled workforce and consumable materials with necessary tools and tackles.

Once the structural design is frozen, resources have to be sourced and allocated for the assignment with commitment of their availability up to the total execution period. Sourcing of materials with necessary tools and tackles also needs to be ensured and confirmed for their uninterrupted availability throughout the execution period.

3. **Activity synchronization.**
   Synchronization of all the activities involved for the assignment is very important with sequencing them so that they are executed without any obstruction and also without causing any uncalled-for safety hazard. Sequencing them is very essential because any alteration in the chronological sequence of the activities can be unsafe for the construction action. Squeezing the time frame attached with any activity – if it is called for, can be carried out under this tool for the ultimate benefit to the project.

4. **Commercial balance between input cost and output gain.**
   It is mandatory to monitor commercial side of any construction activity.
   Flow of funds should be balanced on their input side and outflow direction. If inflow gets obstructed, then the entire economics gets affected with its cascading effects on the overall execution schedule.
   Fund outflow (in form of payment to the suppliers or interest charges) is also equally important. It has

its implications on interest charges (if the funding of the project is outsourced).

5. **External factors.**
   External factors like environmental effects, neighbouring structures, natural calamities, riots etc can affect the construction schedule. They can't be controlled by the construction team and hence they have to be resolved by taking help of local authorities.

6. **Safety aspects.**
   Safety is a mental attitude. It always stands on top of all the activities on field. Construction management is incomplete without adherence to safety.
   The domain of safety covers personal safety as well as safety of the structure under construction.
   It is a fact that ten minor or 'near-mis' accidents lead to one major accident. Hence all out efforts are necessary for plugging all the unsafe acts so that safety matter is perfectly adhered during the construction.
   The result of any unsafe act is chaotic and ends up in loss of time, energy and moral for execution of the work.

7. **Time Management.**
   Time has always remained crux of any activity. Timely actions are bound to give results and if actions have not been taken in time, they can bring only agonies.

   Time factor in construction activities have direct influence on economics of the assignment. Any delay in executing one or more activities ends up in delay in putting the unit (under construction) to use which is direct loss of production / remuneration expected from the unit. Delay in execution period also affects the execution cost where more than estimated manpower are consumed in execution at higher cost of consumable materials for execution.

Acid alkali resistant treatment with carbon bricks laid in suitable mortar.

It was understood in last quarter of year 2023 that a special acid alkali resistant treatment was essential in one of large size steel manufacturing industries in India. The facts of the case and further details of the subject are as placed below.

A chemical reaction between H_2SO_4(with 98% concentration) and $Ca(OH)_2$(with more than 15 pH) which was a part of manufacturing process of a patented product, was taken up in a reactor having 8 mm thick plate thickness. The reaction was such that it would generate 325^0 C heat for 20 minutes for 3 times in a day. It was found to be creating severe corrosive environment. Heat of more than 300^0 C aggravated the corrosive effects, as a result of which the reactor with MS plate of 8 mm thickness had developed many holes on its body. Sprinkling of water in-between the reaction (which was an operational requirement) was further aggravating the severity of corrosion.

The entire process was closely examined and it could be concluded that thermal shock was equally responsible for severe corrosive condition with the same severity as chemical reaction. Many options were examined to arrive at a successful solution to the challenging situation.

Subsequent to exploration of all options, carbon brick lining was found to be the only protective treatment for the situation. Mortar for this treatment was another challenge because although carbon bricks would resist the chemical as well as thermal shock but if mortar is unable to absorb that combined shock, then the ultimate total success would not be achieved. The mortar with carbon fill, graphite fill and other fillers was tried out in a laboratory scale experiment. Both carbon filled and graphite filled mortar survived the condition for 48 hours and then the treatment started deterioration. The mortar with very exclusive special fillers had given an ultimate success.

The laboratory scale trial pot has been tested at the industry and it is confirmed that the reaction carried out in the trial pot, didn't deteriorate the metal plates and accordingly, the treatment is 100% successful in all the parameters as desired by the industry.

The entire mission was very nicely coordinated with the experts of the field for their Technical support as well as for preparation of trial pot lined with the said treatment.

Purpose of sharing this information is to emphasise the potential of acid alkali resistant treatment with carbon bricks laid in tailor made mortar to suit the actual field requirement.

Smiles …. and structural retrofit …

It may sound a misfit combo of words at the first instance, but this is meaningful combo of words resulting into emotional glue for the ultimate success of the activity.

Structural retrofit or structural rehabilitation of structural refurbishment is necessitated when the structure is found either underperforming or needs some functional augmentation.

Smile while conceptualising:

Very conceptualising of the fact of structural retrofit; needs to be accepted with a sweet smile for all the team members because then that is carried almost throughout the next set of actions. Smile holds a symbol here of the fact that 'everyone is on same page'.

Smile while shaping up the concept on software or drawing board:

Smile holds a very specific function during this particular activity. It just acts like a catalyst during the process of giving shape to the concept of structural retrofit. The action is very similar to talking to each member demanding retrofit and consoling it by ensuring that each one gets desired enhancement of functional life.

Smile during execution:

Execution of structural retrofit action is like a surgery on the member. It is indeed painful but just as doctors do it with some anaesthesia and some 'healing touch' for pain relief, an innocent smile at the member while executing the painful activities like scrapping etc would certainly get a response from the member. Please try it.

Smile at the end:

When the member is ready to be functional after getting retrofitted, the energy, the spirit and the enthusiasm experienced by the member always overflows from the (sur)face of the member.

The brighter it looks as compared to before the retrofit, the more powerful it becomes compared to before the retrofit and the more sustainable it becomes compared to before retrofit.

Well dressed (sur)face with freshly applied paint finish is nothing less than a smile on the (sur)face of the member.

The journey of structural retrofit which began with

"Smiles ... and ... structural retrofit",

also ends with

"Structural retrofit .... and ... Smiles".